Shells Are Skeletons

People walk around with their skeletons inside them. But there are many amazing creatures which wear their skeletons on the outside. Tiny snails that crawl on the plants in your garden, giant oysters deep under the sea, and hundreds of other members of the mollusk family all live snugly inside their shells. The shell keeps them safe from their enemies, and as they grow the shell gets bigger too.

In beautiful illustrations and easy-to-read text, Joan Victor explains how shells are formed, and how the creatures inside them live and move and eat. Her book is an exciting introduction to the fascinating variety of shells.

Shells Are Skeletons

Written and illustrated by Joan Berg Victor

Thomas Y. Crowell Company   New York

LET'S-READ-AND-FIND-OUT SCIENCE BOOKS

LIVING THINGS: PLANTS

Corn Is Maize: The Gift of the Indians
Down Come the Leaves
How a Seed Grows
Mushrooms and Molds
Plants in Winter
Roots Are Food Finders
Seeds by Wind and Water
The Sunlit Sea
A Tree Is a Plant
Water Plants
Where Does Your Garden Grow?

LIVING THINGS: ANIMALS, BIRDS, FISH, INSECTS, ETC.

Animals in Winter
Back Where They Came From: The Eels' Strange Journey
Bats in the Dark
Bees and Beelines
Big Tracks, Little Tracks
Birds at Night
Birds Eat and Eat and Eat
Bird Talk
The Blue Whale
Camels: Ships of the Desert
Cockroaches: Here, There, and Everywhere
Corals
Ducks Don't Get Wet
The Emperor Penguins
Fireflies in the Night
Giraffes at Home
Green Grass and White Milk
Green Turtle Mysteries
Hummingbirds in the Garden
Hungry Sharks
It's Nesting Time
Ladybug, Ladybug, Fly Away Home
The Long-Lost Coelacanth and Other Living Fossils
My Daddy Longlegs
My Visit to the Dinosaurs
Opossum
Sandpipers
Shells Are Skeletons
Shrimps
Spider Silk
Spring Peepers
Starfish
Twist, Wiggle, and Squirm: A Book About Earthworms
Watch Honeybees with Me
What I Like About Toads
Why Frogs Are Wet

THE HUMAN BODY

A Baby Starts to Grow
Before You Were a Baby
A Drop of Blood
Fat and Skinny
Find Out by Touching
Follow Your Nose
Hear Your Heart
How Many Teeth?
How You Talk
In the Night
*Look at Your Eyes**
My Five Senses
My Hands
The Skeleton Inside You
Sleep Is for Everyone
*Straight Hair, Curly Hair**
Use Your Brain
What Happens to a Hamburger
*Your Skin and Mine**

And other books on AIR, WATER, AND WEATHER; THE EARTH AND ITS COMPOSITION; ASTRONOMY AND SPACE; and MATTER AND ENERGY

**Available in Spanish*

Published simultaneously in Canada by Fitzhenry & Whiteside Limited, Toronto. Manufactured in the United States of America.

Library of Congress Cataloging in Publication Data Victor, Joan Berg. Shells are skeletons. SUMMARY: Describes how mollusks, such as clams, oysters, limpets, and snails, build their shells and use them. 1. Mollusks—Juv. lit. [1. Mollusks. 2. Shells] I. Title.
QL405.2.V5 594'.04'7 75-23258 ISBN 0-690-01038-9 (CQR)

1 2 3 4 5 6 7 8 9 10

Shells Are Skeletons

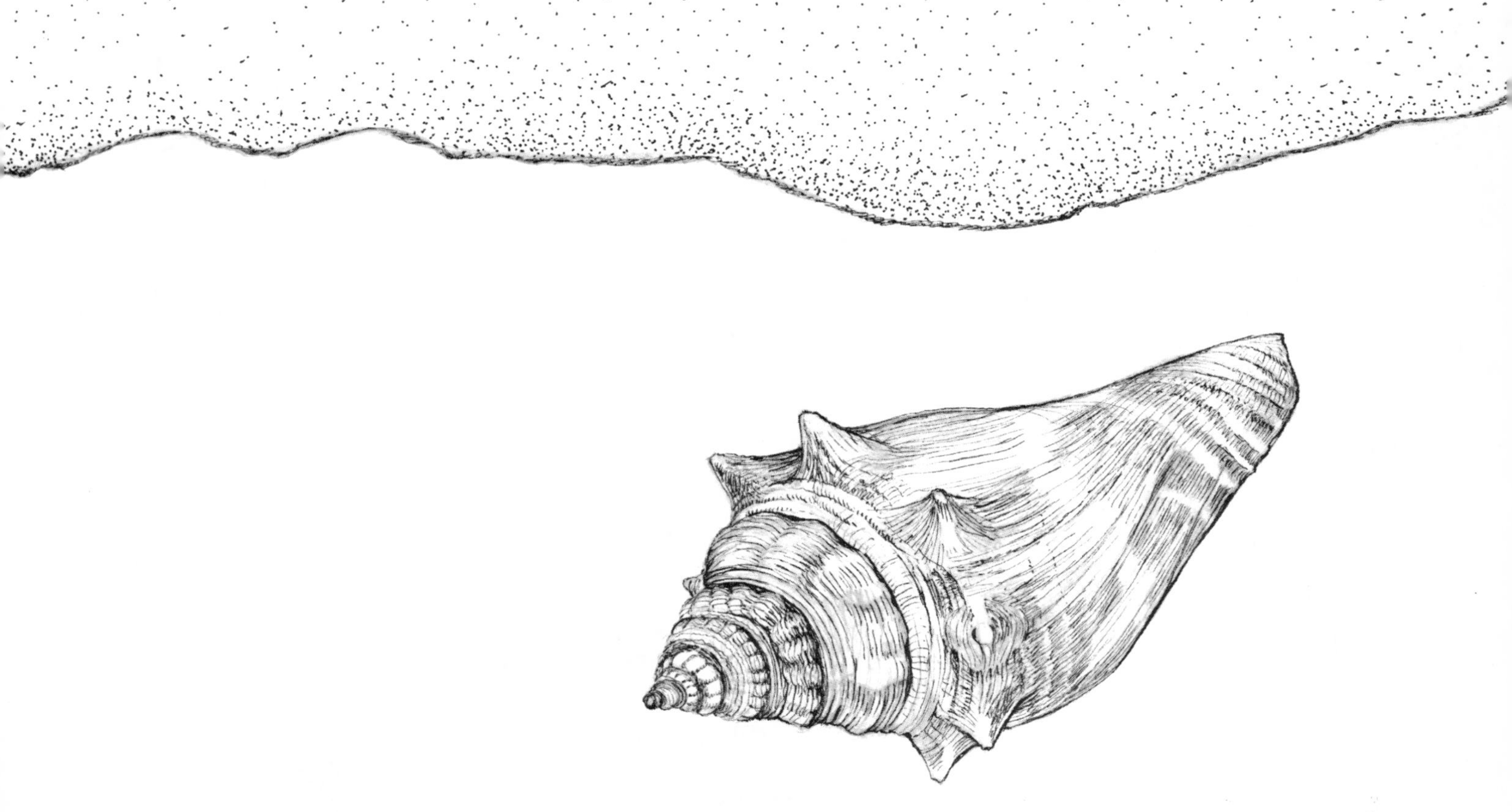

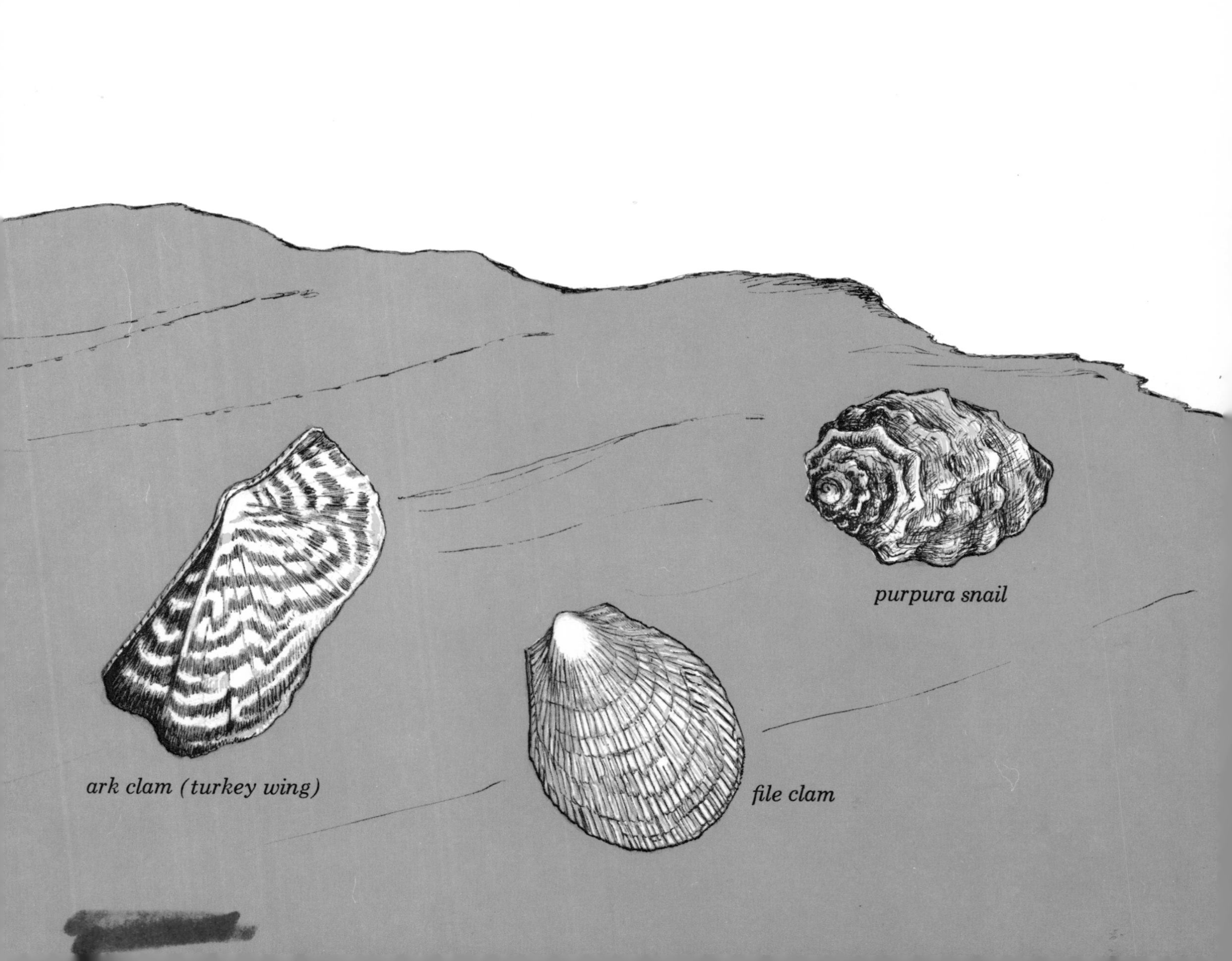
purpura snail
ark clam (turkey wing)
file clam

We walk around with our skeletons inside our skin. Nobody can see our bones. But some creatures wear their skeletons on the outside. That's what mollusks do. A mollusk's skeleton is a shell. Snails, clams, oysters, scallops, and lots of other animals are mollusks. Most mollusks live in the oceans and in ponds and in rivers. Some mollusks live on land.

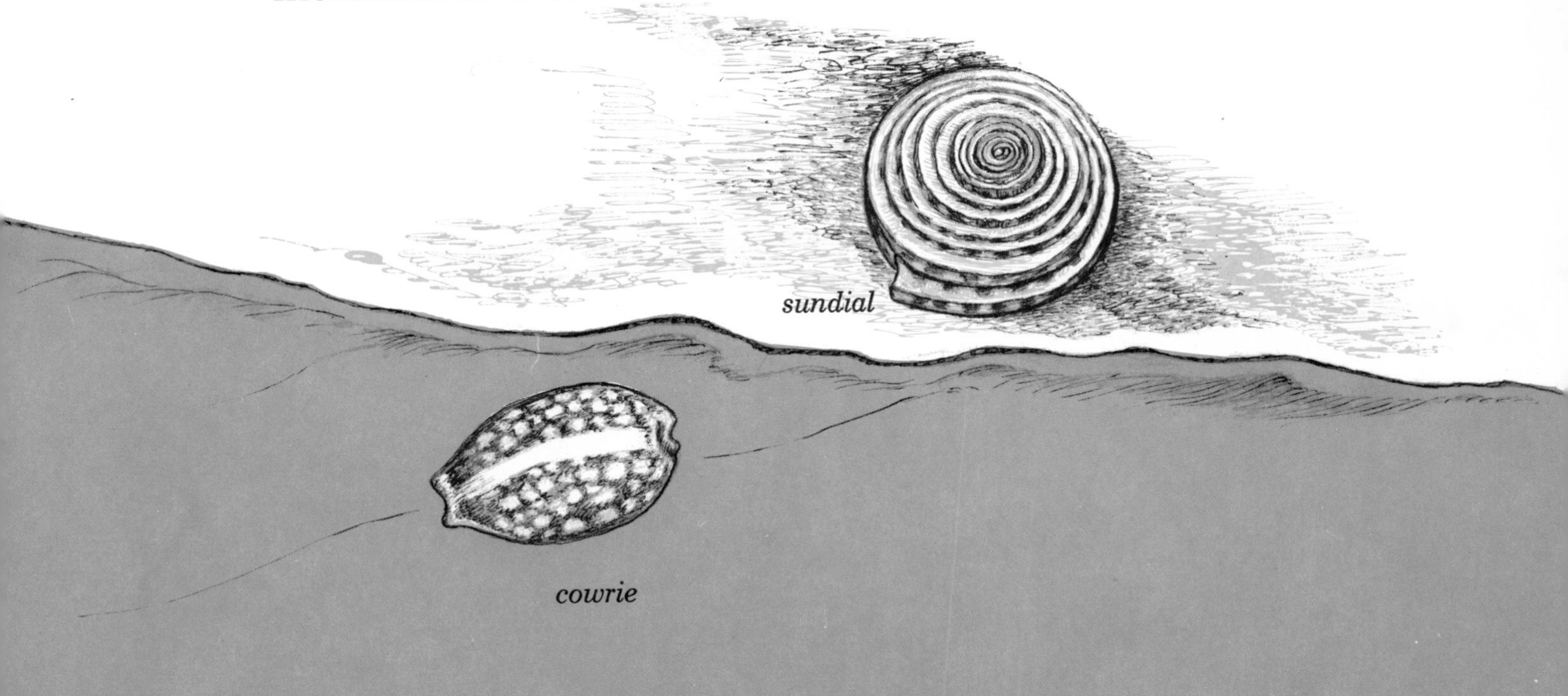

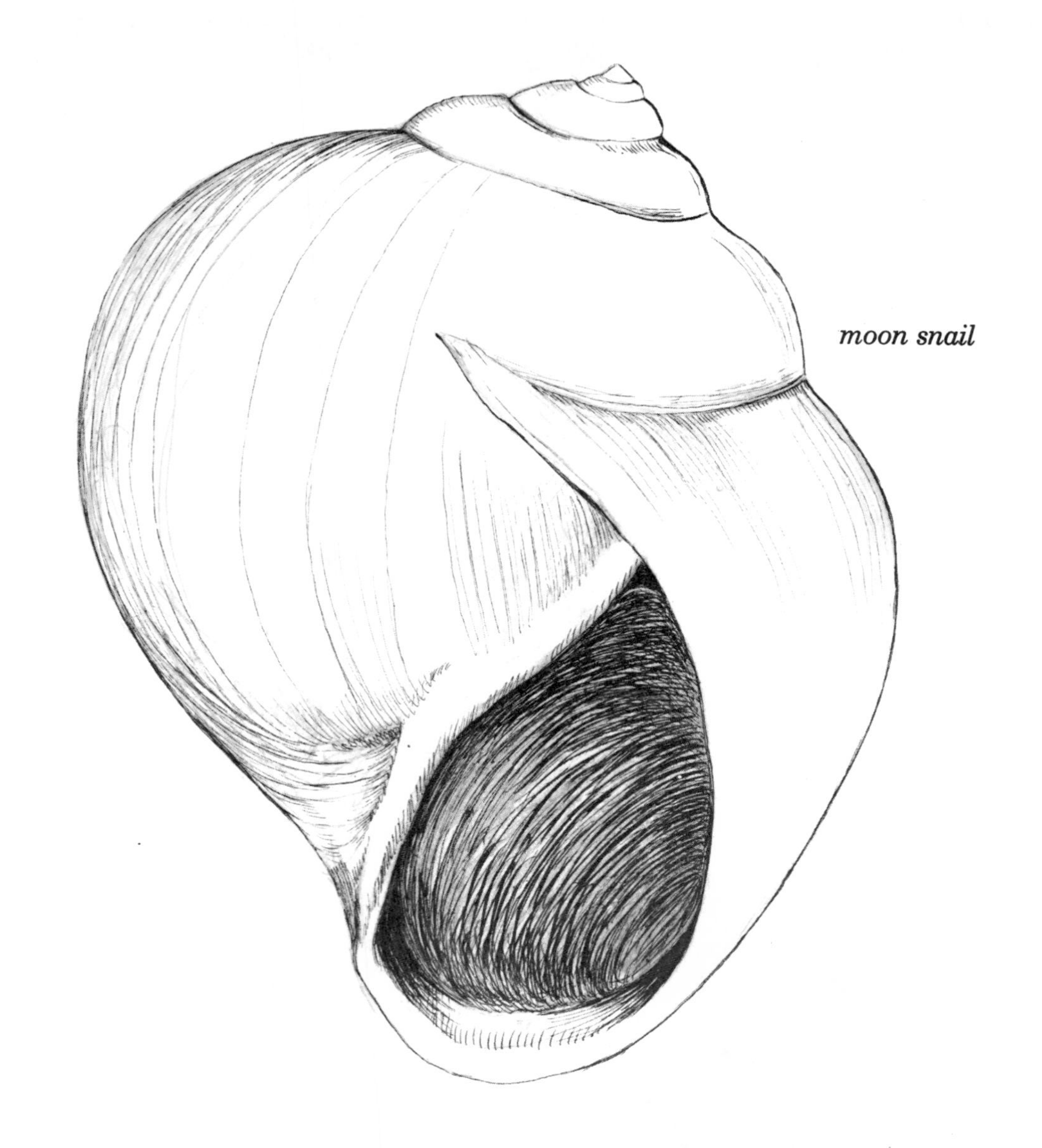

The shell of the mollusk helps protect the creature inside from other animals that might eat it.

A cloak-like layer of skin covers the sides and back of the mollusk's body. This layer is called the mantle. The mantle is covered with special cells. Lime, a mineral, comes out of these cells. It makes tiny layers of shell.

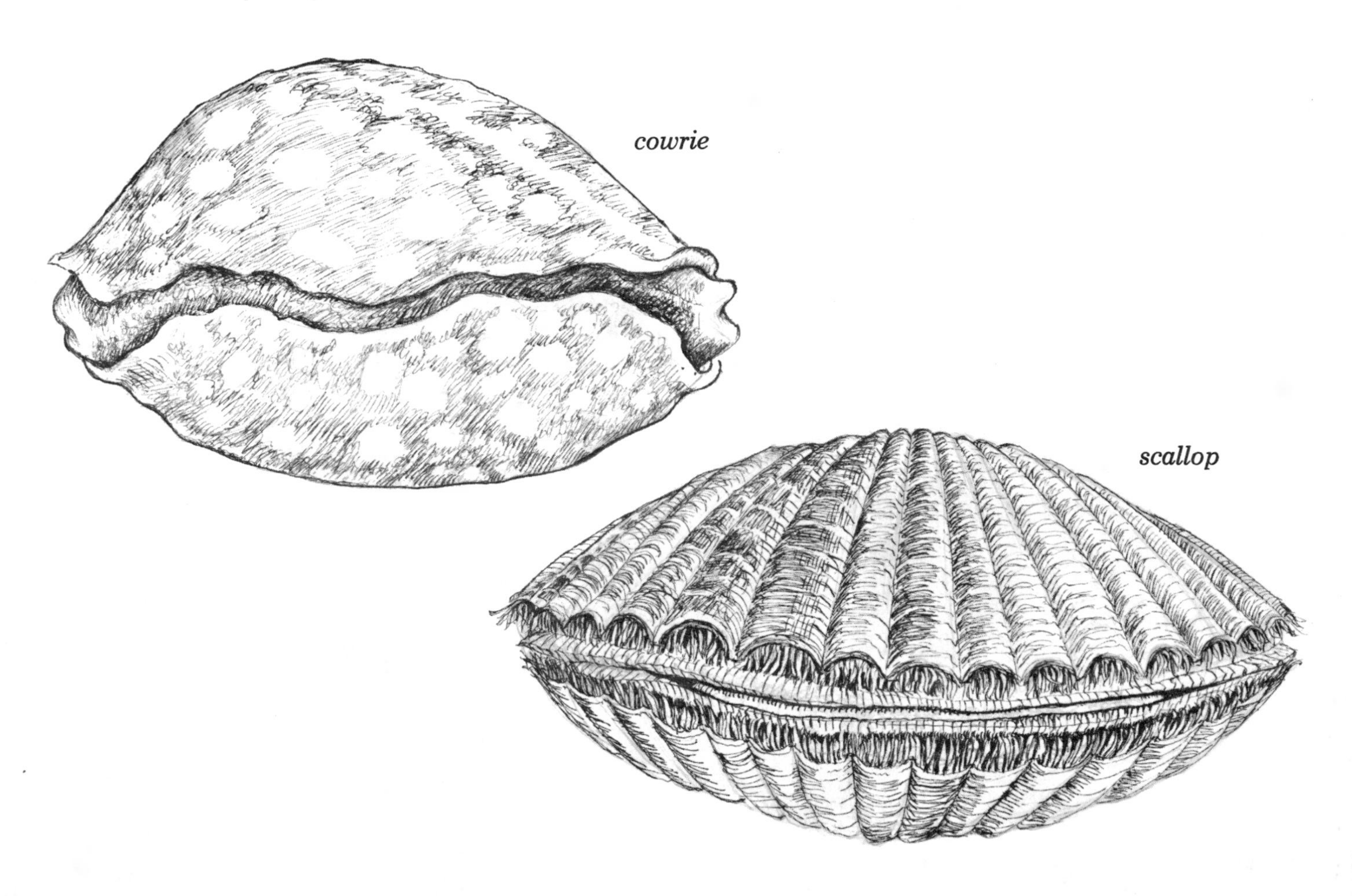

Several kinds of limy layers are formed. Some of these are laid crisscross to those above and below. Sometimes the layers are colored. When the animal's mantle stops building its shell, the animal inside cannot grow larger.

microscopic view of growing edge of outer lip of purpura snail

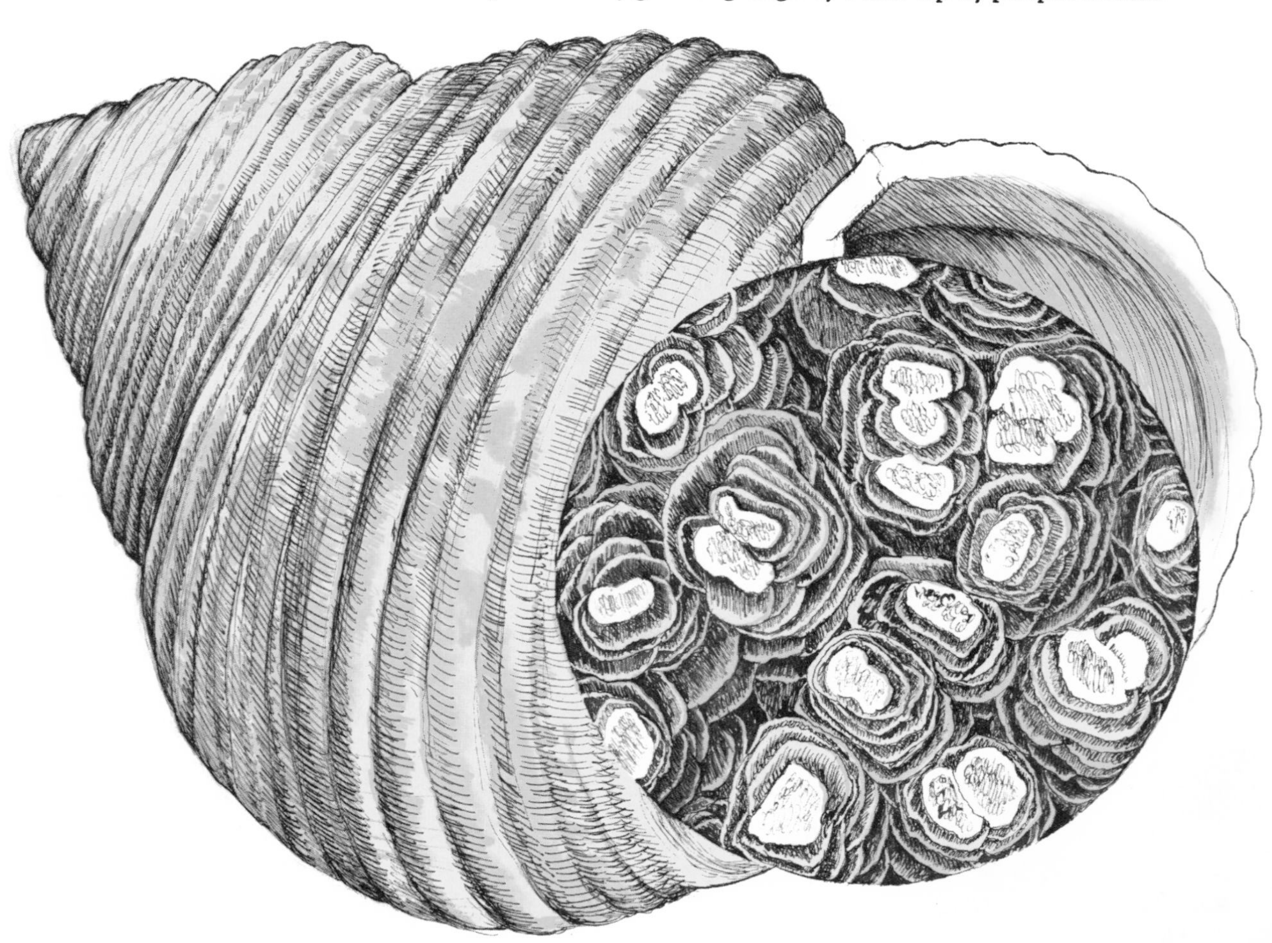

Inside the shell some mollusks add layers that lap over one another like shingles on a roof. When you open the shell you see these smooth, shiny layers. Light catches the edges of the layers and makes colors that seem to change. These layers are called mother-of-pearl.

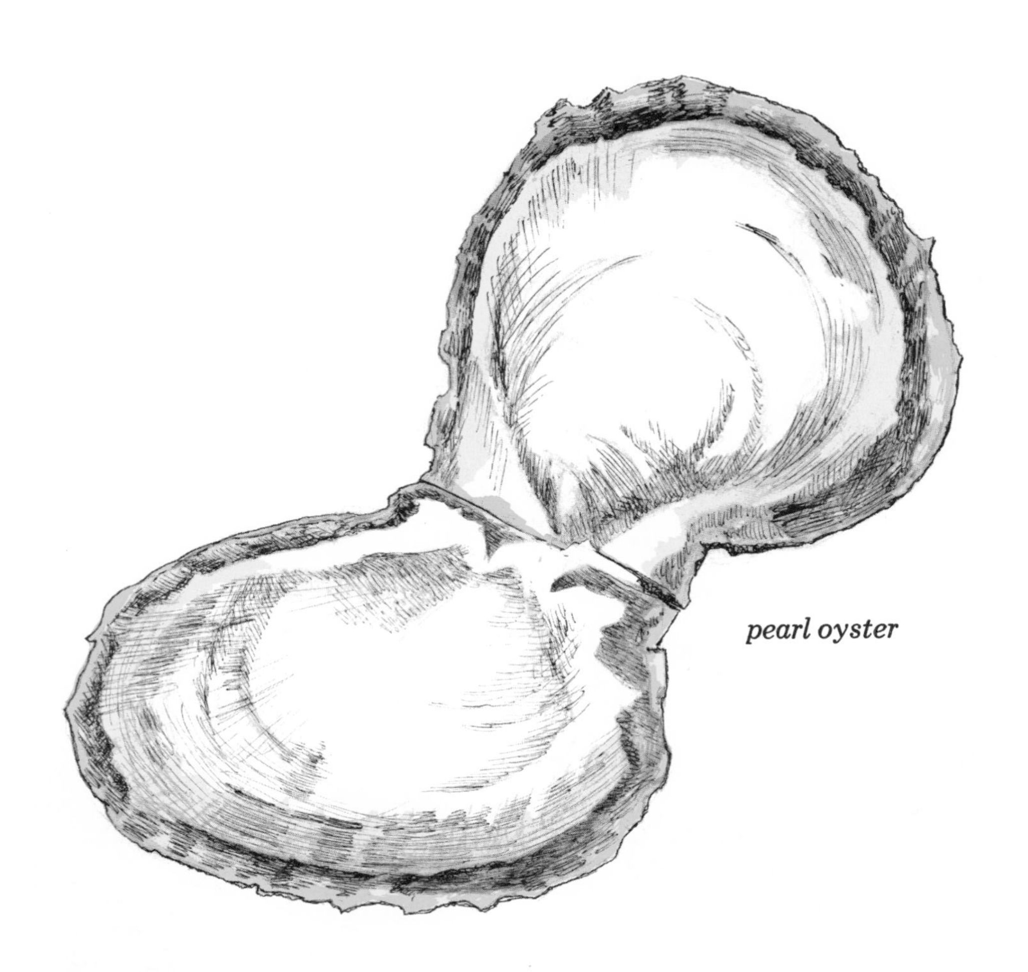

pearl oyster

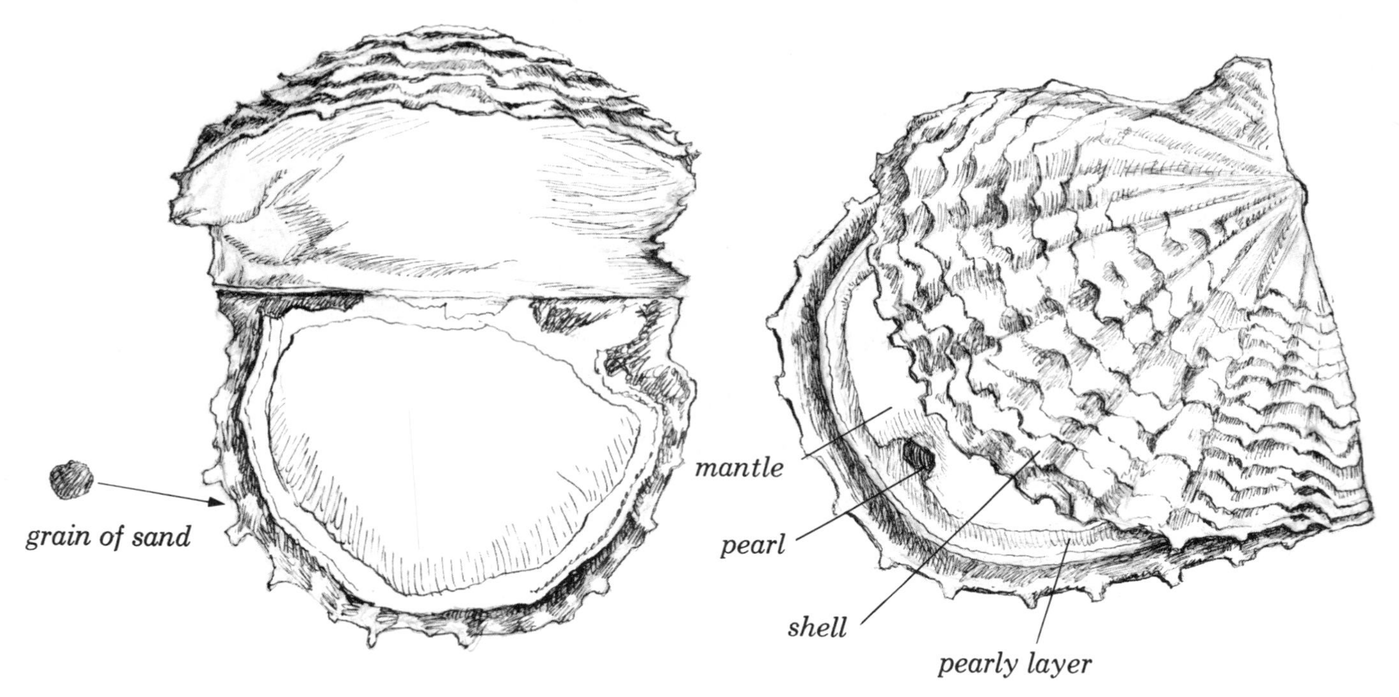

Sometimes a grain of sand or a bit of broken shell gets inside a shell. It gets covered with layers of mother-of-pearl, so that it will be smooth and round. When it has many of these layers, it is a real pearl. Sometime you might find a pearl in an oyster. Some pearls are found by divers, in giant oysters deep down at the bottom of the ocean. There are even oyster farms where pearls are grown. All clams and some snails may build pearls of various kinds.

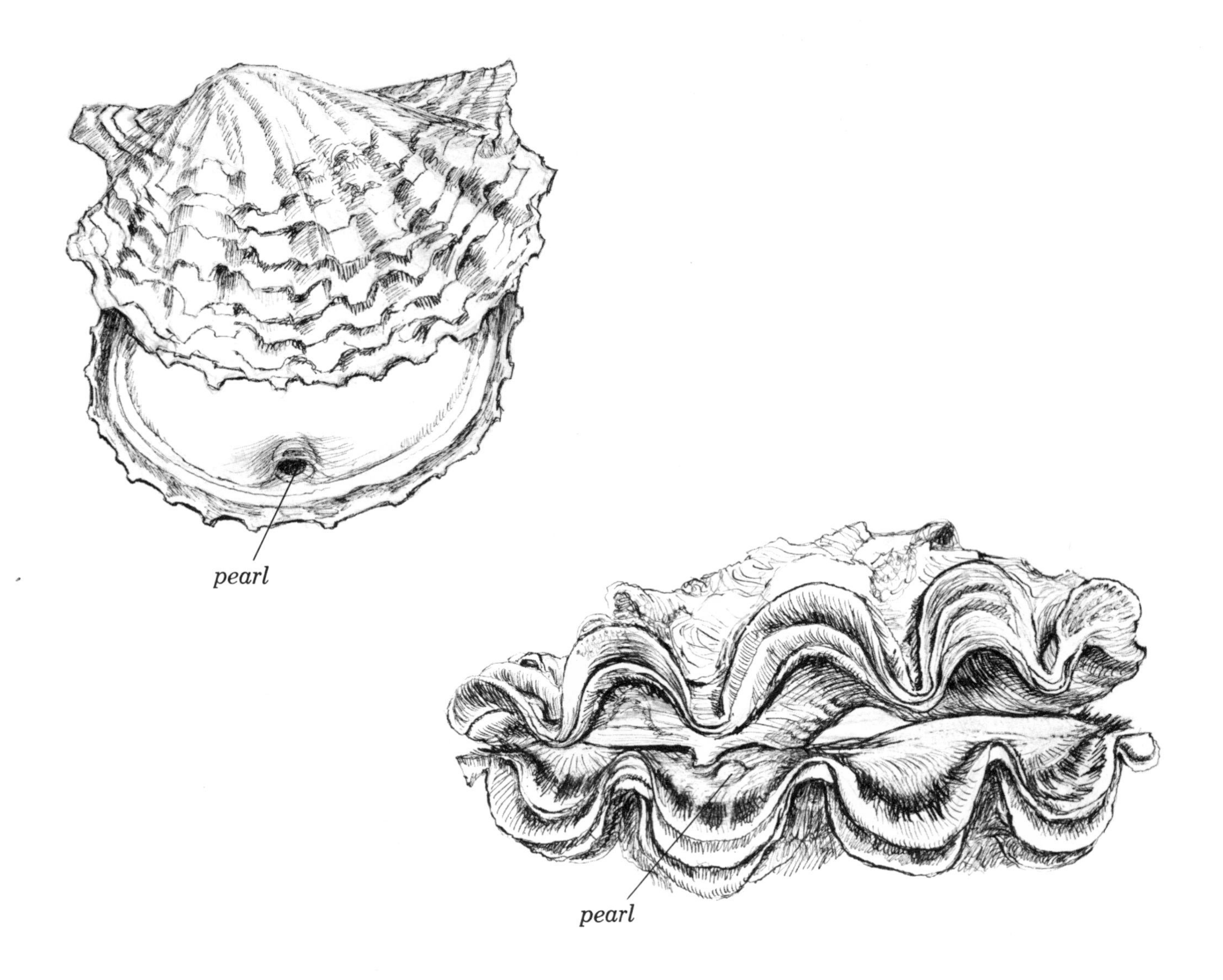
pearl
pearl

In some mollusks a dull brown, rough covering hides the color and markings on the outside of the shell. When the mollusk dies, sand and water wear away this outer, skin-like covering.

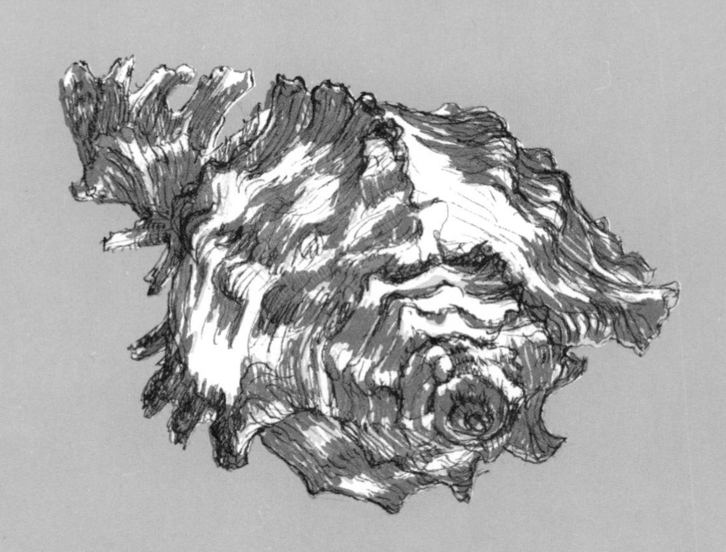

By the time you find the shell, the skin may be cleaned off.

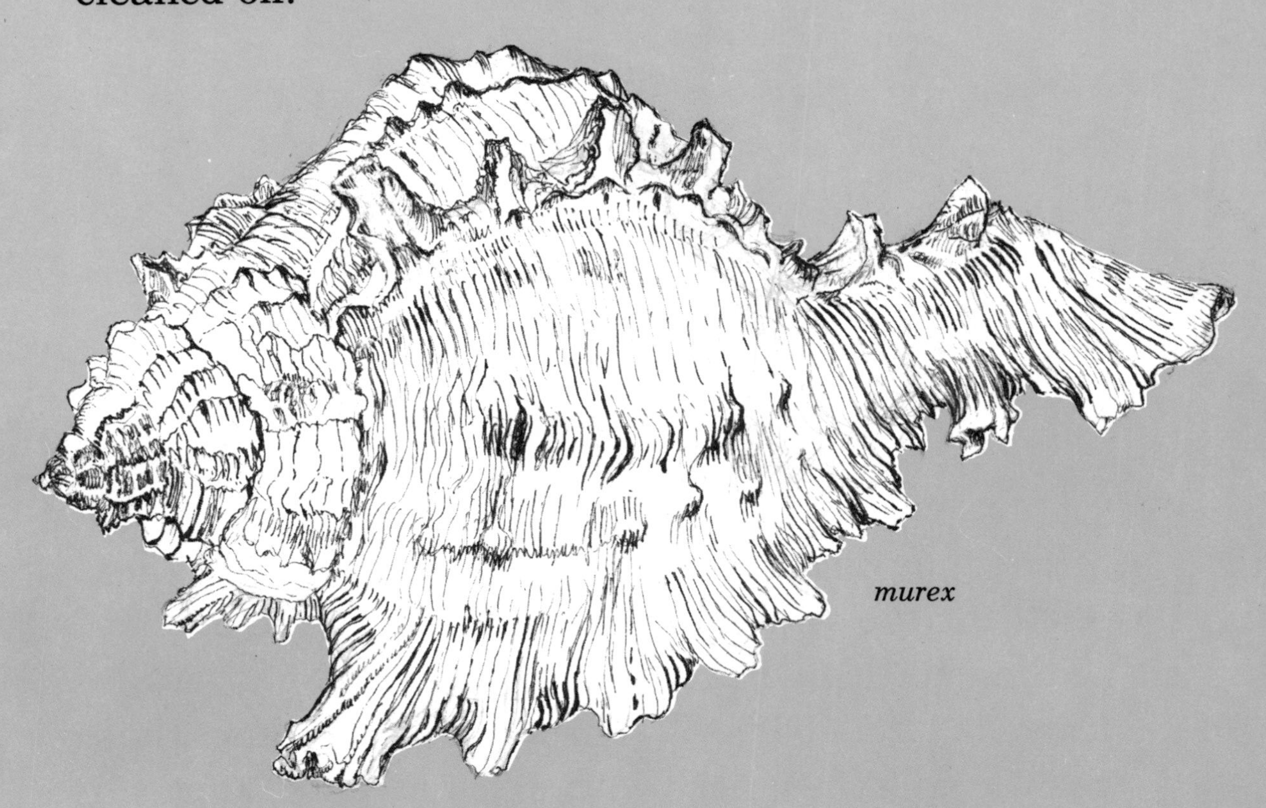

murex

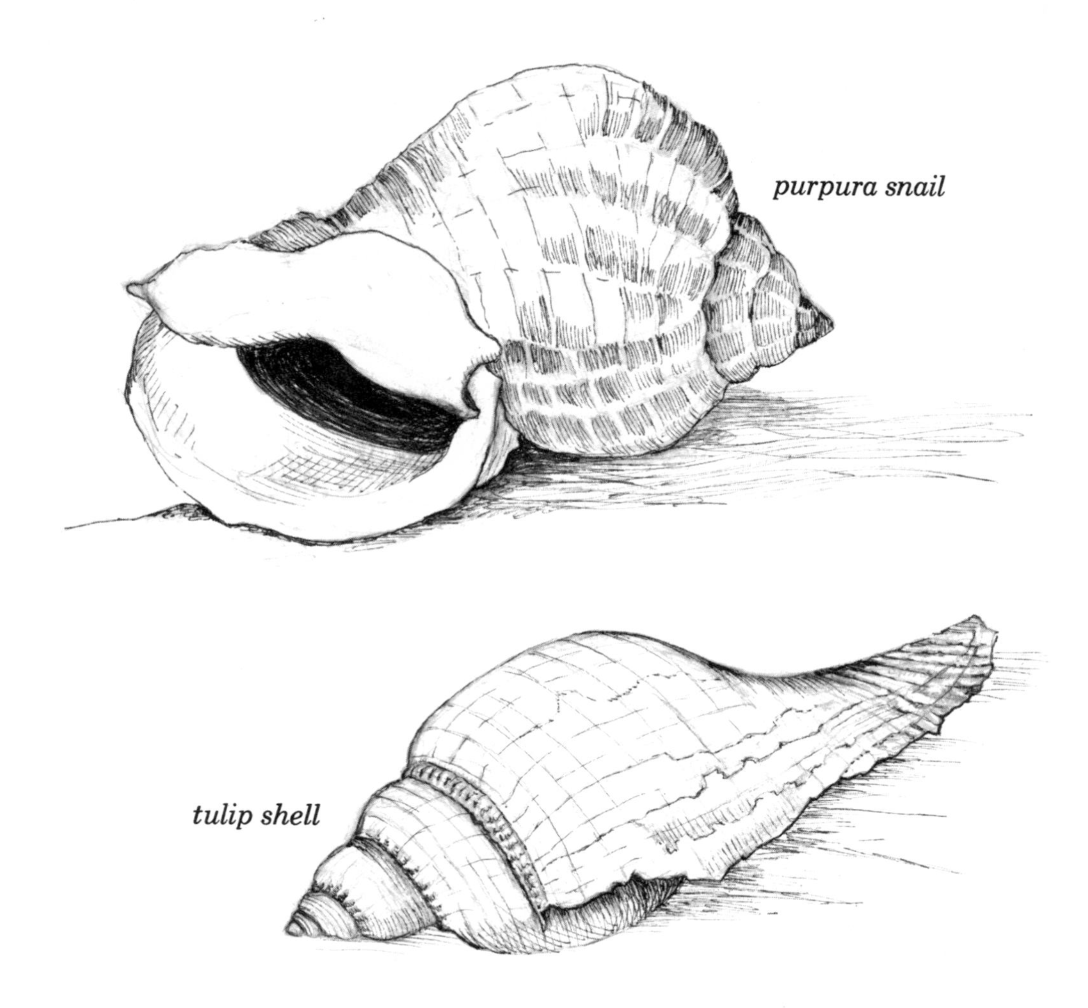

The mantle spreads over the outside edge of the shell. As the animal grows, the cells on the outside of the mantle add lime to the edges of the layers. The shell gets bigger. If the shell gets broken, the mantle fills the cracks with extra layers of lime.

Shells begin to grow when the new animal is still in its egg. In most mollusks, this early shell is not shed. It is just added to as the mollusk grows.

neptune snail egg capsules with young hatching

Sometime you may find the egg case of a mollusk called a sea snail. If you look carefully through the egg case, you may see hundreds of tiny snails. Their shells are not much larger than a grain of sand. Each tiny snail may grow to be eight or ten inches long.

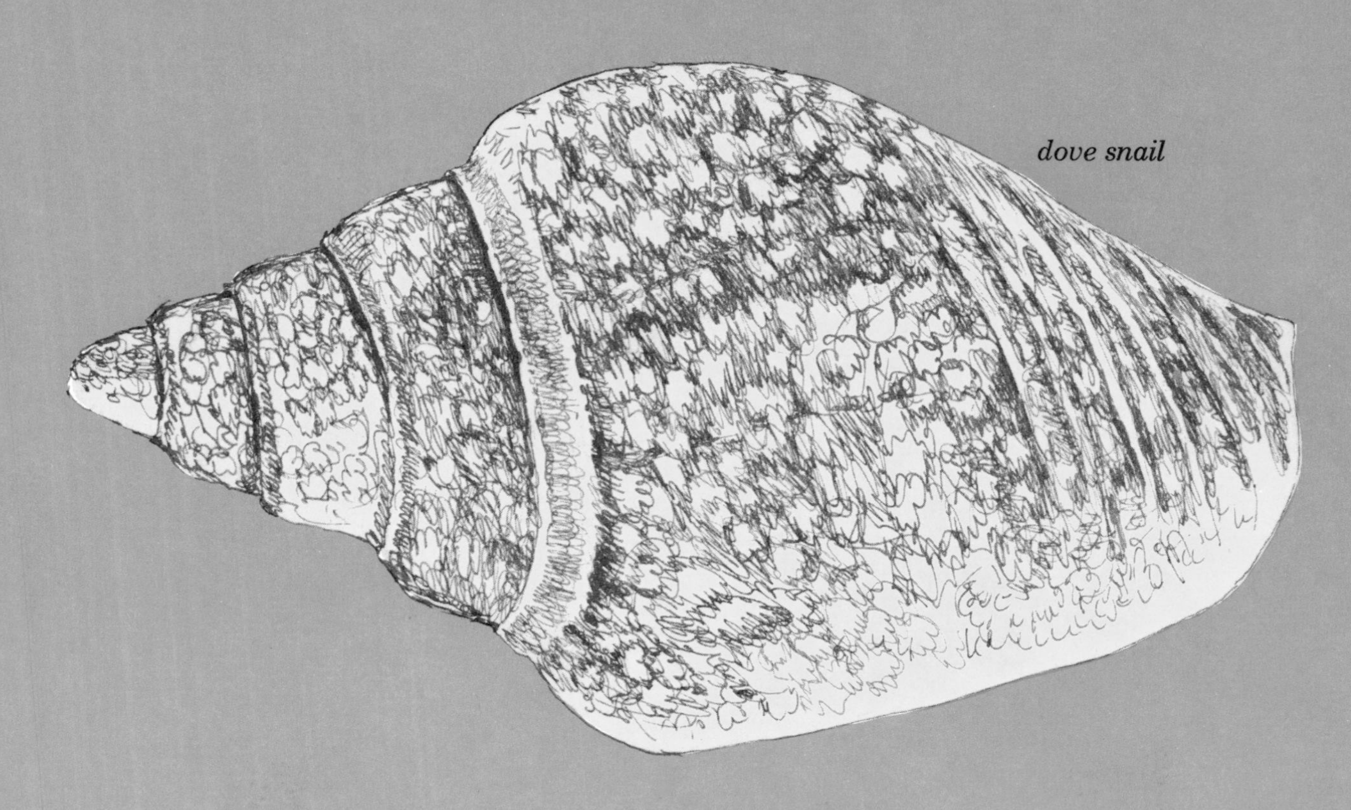

dove snail

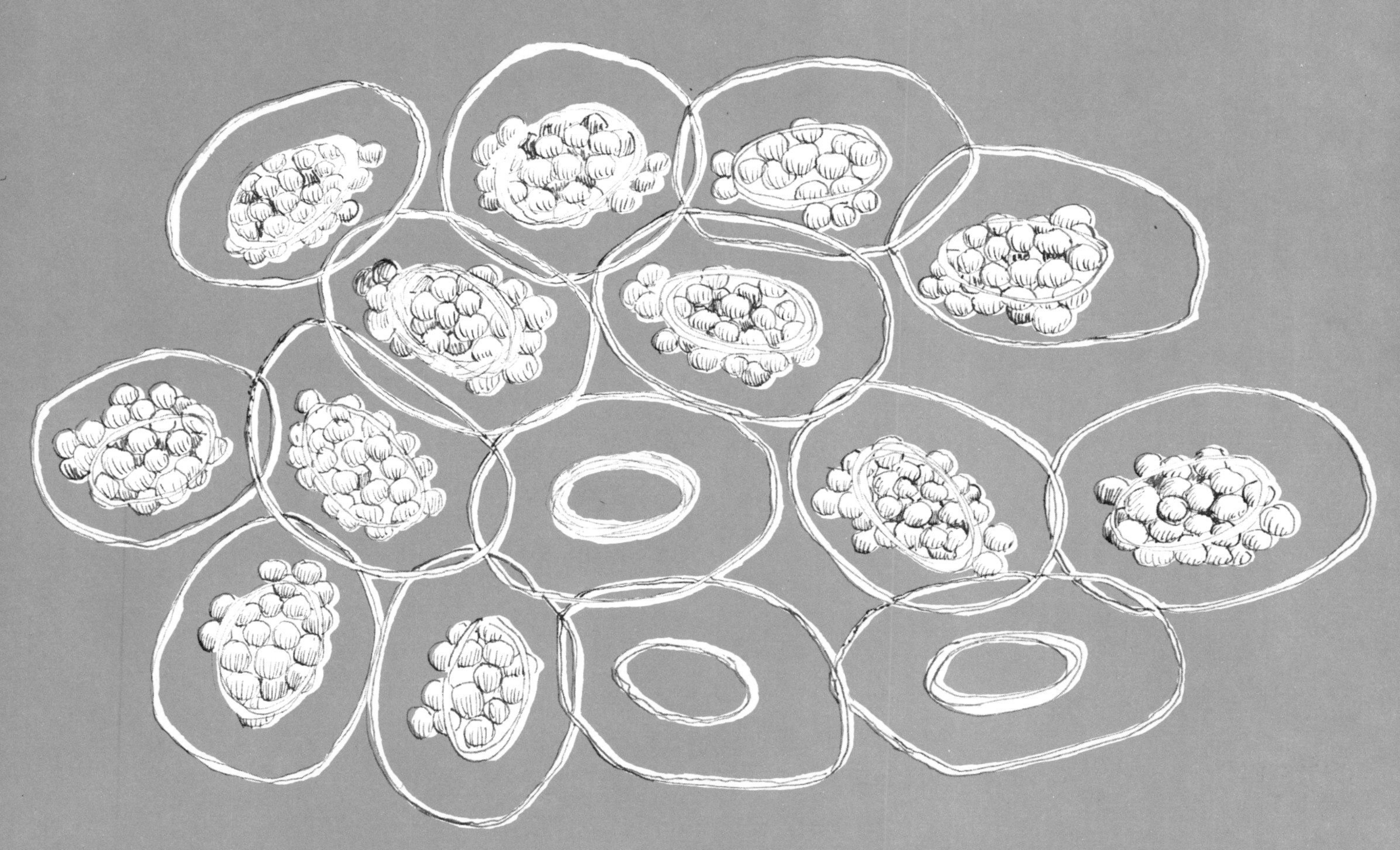

jelly capsules containing eggs of dove snail. In each capsule there may be more than a dozen eggs. Snails have already hatched from the empty capsules.

The shell is made of lime. Lime is a mineral. It is also called calcium carbonate. Your teeth are made of it; so are your bones. You get calcium carbonate from fruits and vegetables you eat, and from milk you drink. You may get some from water, too.

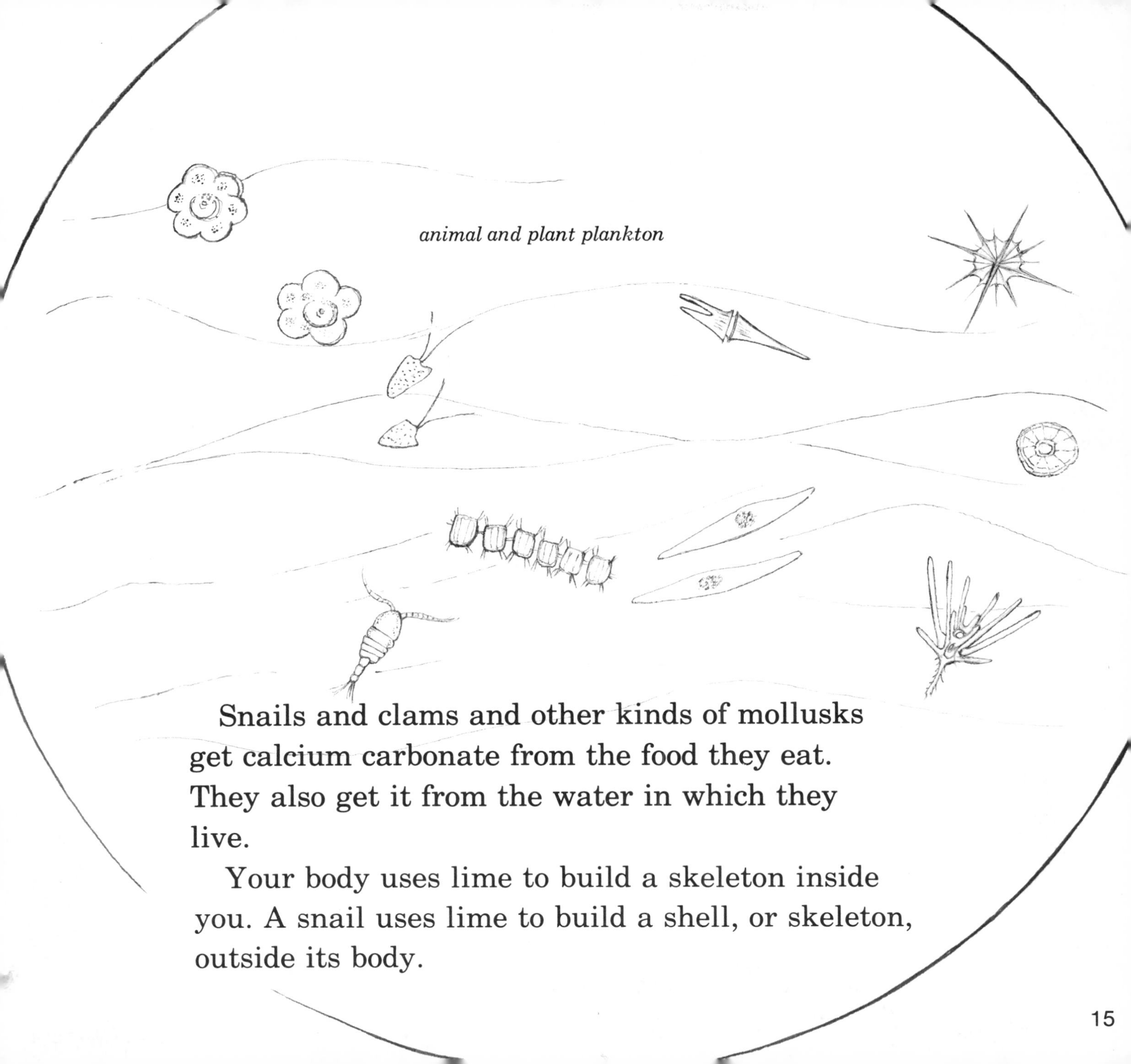

Snails and clams and other kinds of mollusks get calcium carbonate from the food they eat. They also get it from the water in which they live.

Your body uses lime to build a skeleton inside you. A snail uses lime to build a shell, or skeleton, outside its body.

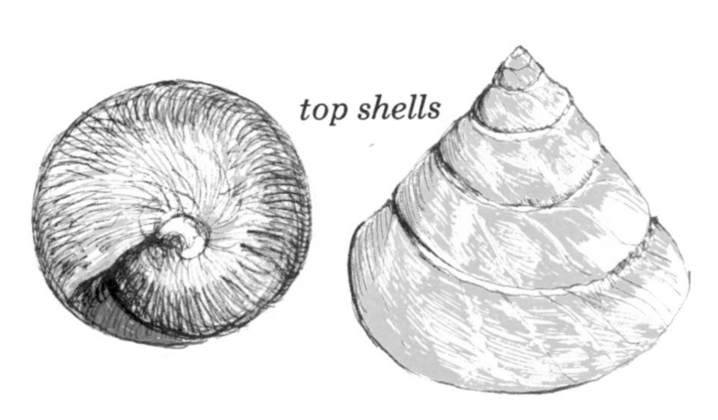

top shells

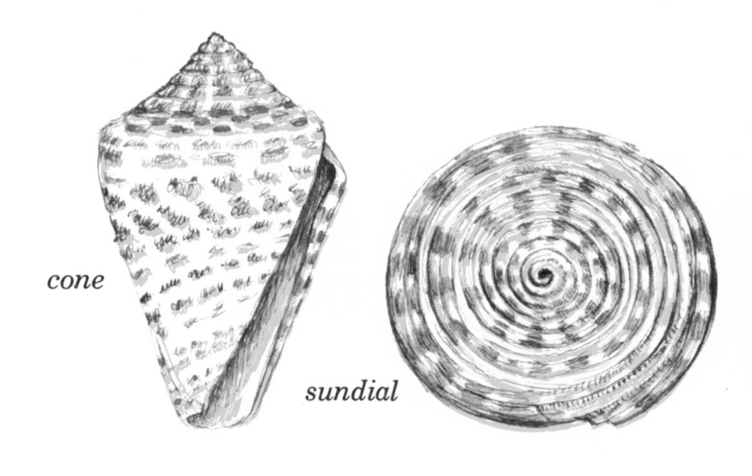

cone

sundial

shiny auger

Some mollusks have only a single shell. They are called univalves. *Uni* means one, and *valve* is a word for shell. Periwinkles, snails, and limpets all have one shell.

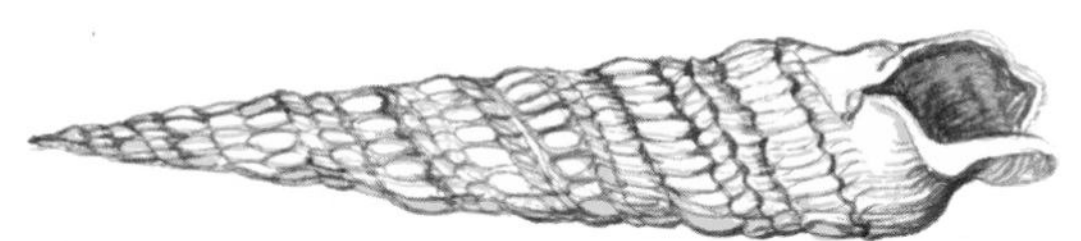

Florida auger

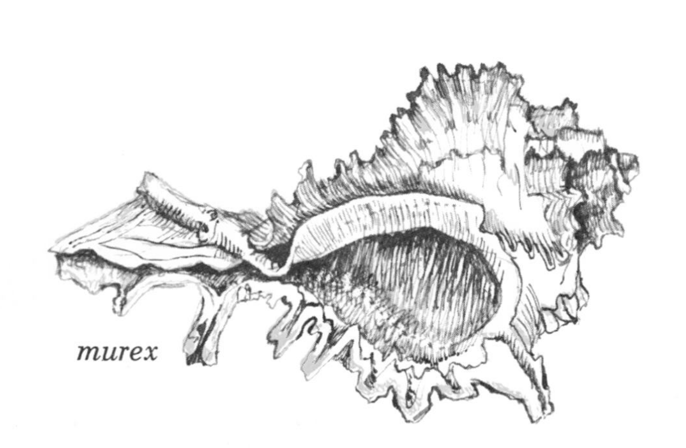

murex

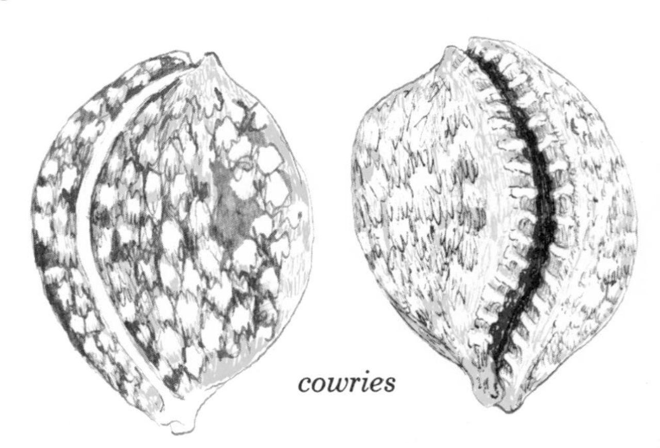

cowries

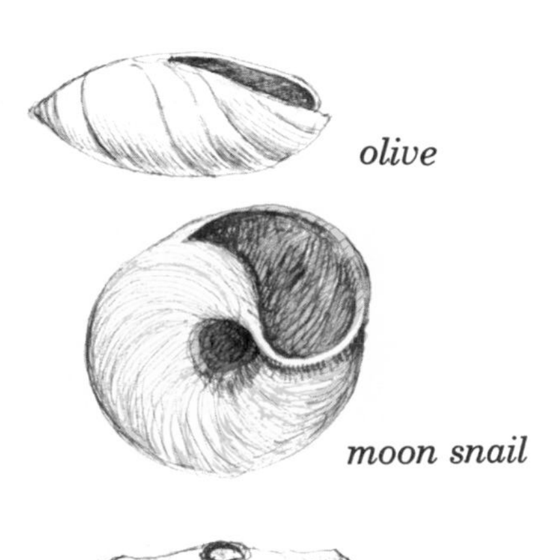

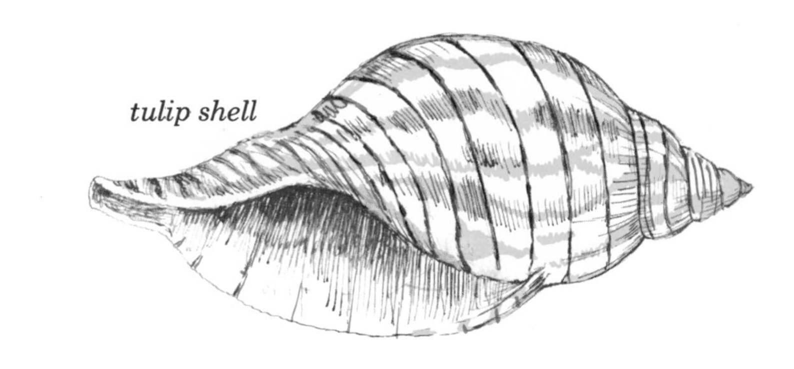

keyhole limpet

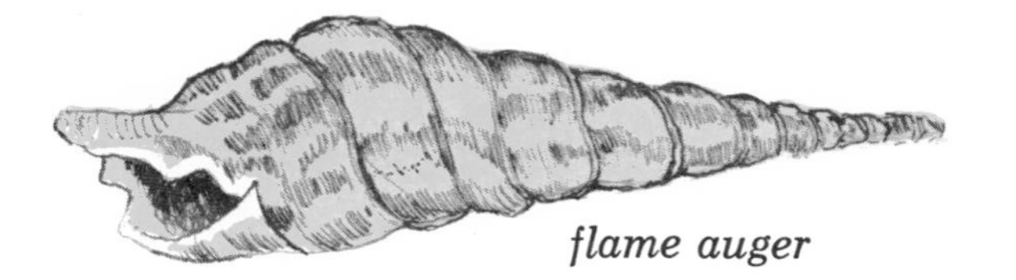

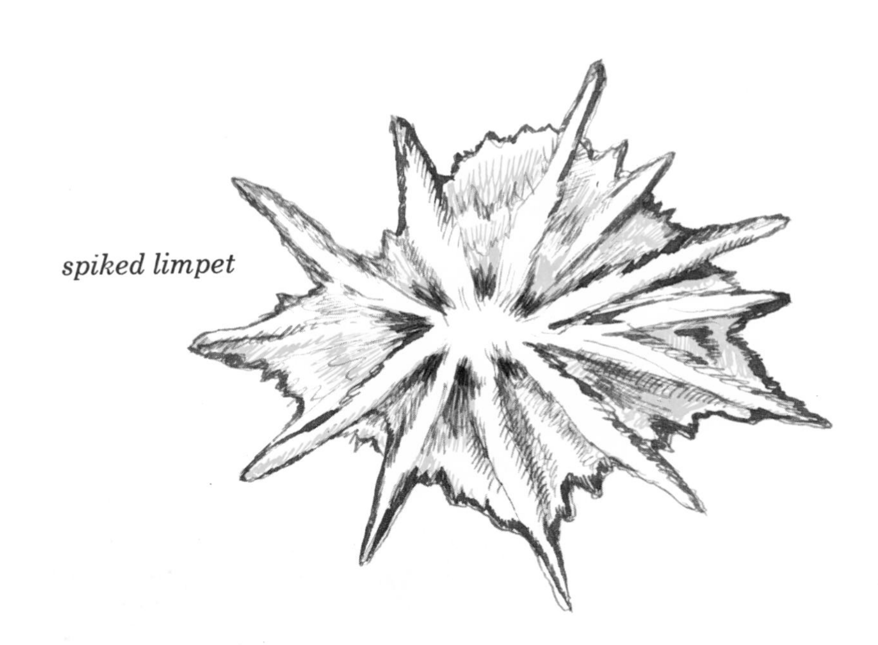

When a snail is feeding, it comes part way out of its shell. Its flat foot holds it to a rock or a branch. If a crab that might eat it comes along, the snail quickly pulls head and foot into the shell. Then the snail closes up the opening to the shell.

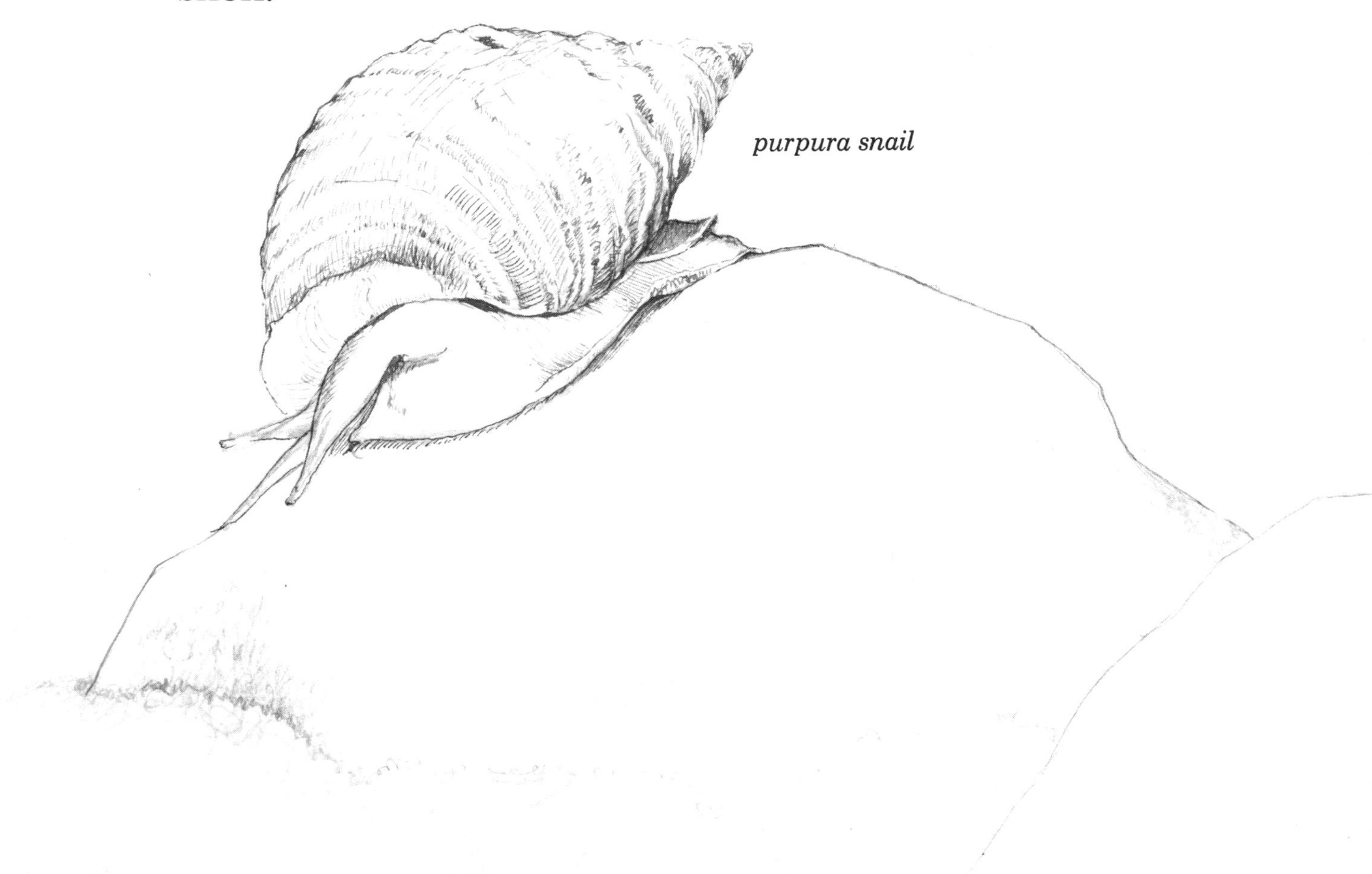

purpura snail

Another kind of mollusk has two shells hinged together. These mollusks are called bivalves. There are about ten thousand kinds of bivalves. Clams, oysters, and scallops have two shells.

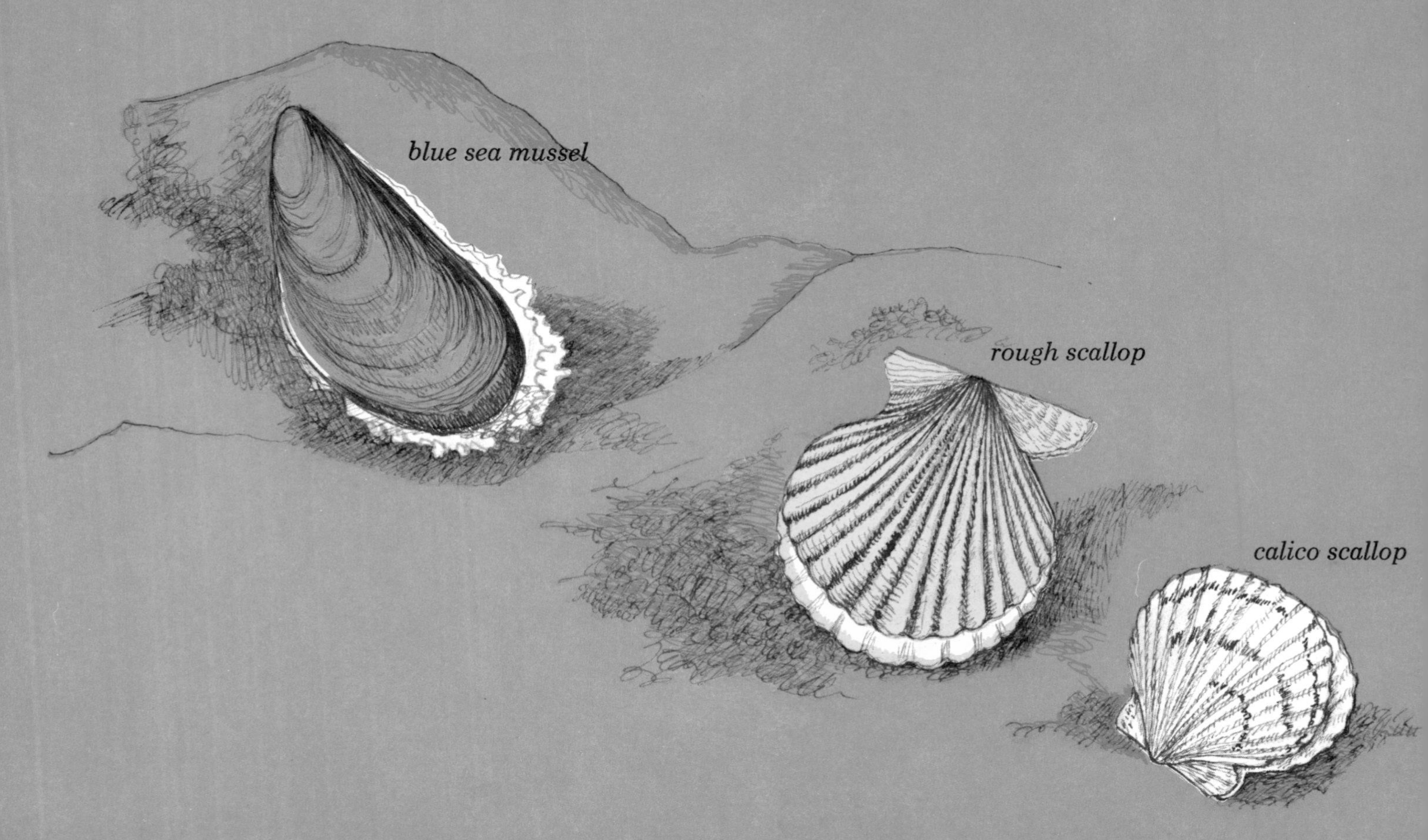

Pacific thorny oyster

rough file clam

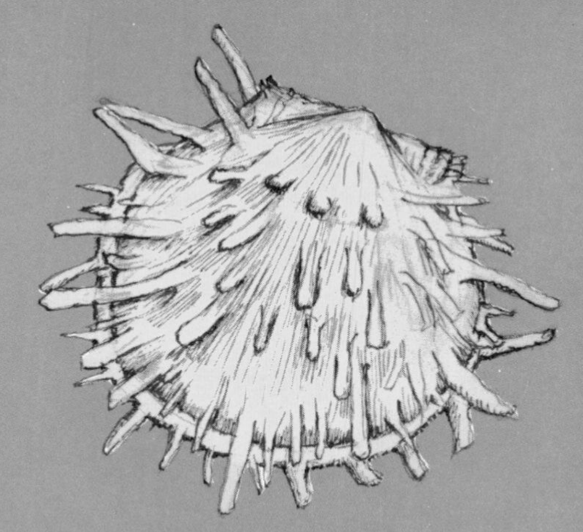

Atlantic thorny oyster

king venus

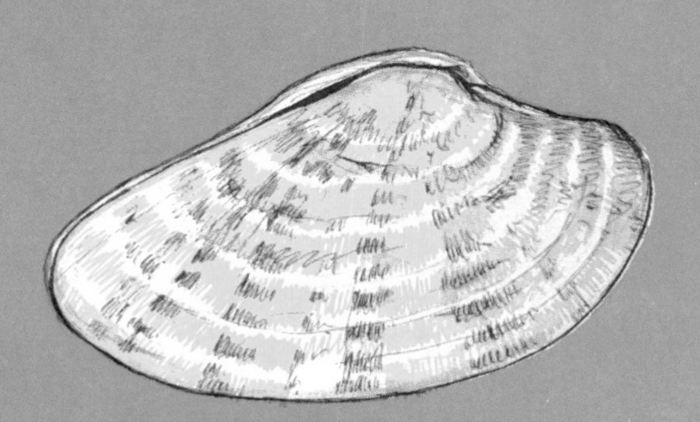

sunray venus

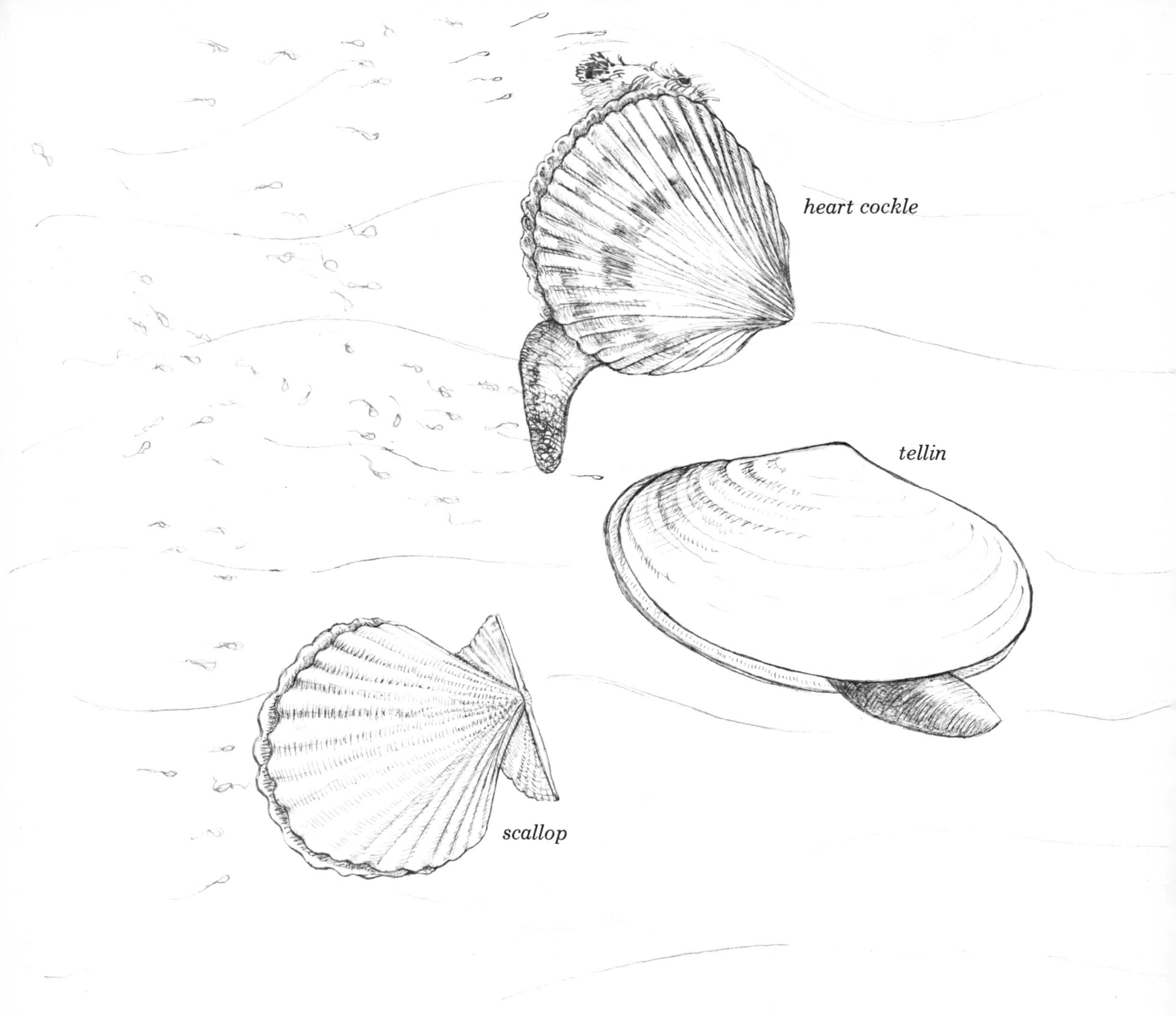
heart cockle
tellin
scallop

In the water, the shells open a little bit. Water washes between the shells and over the mouth of the animal. The water carries tiny plants and animals and minerals. That's the way the bivalves get their food and the lime for their shells.

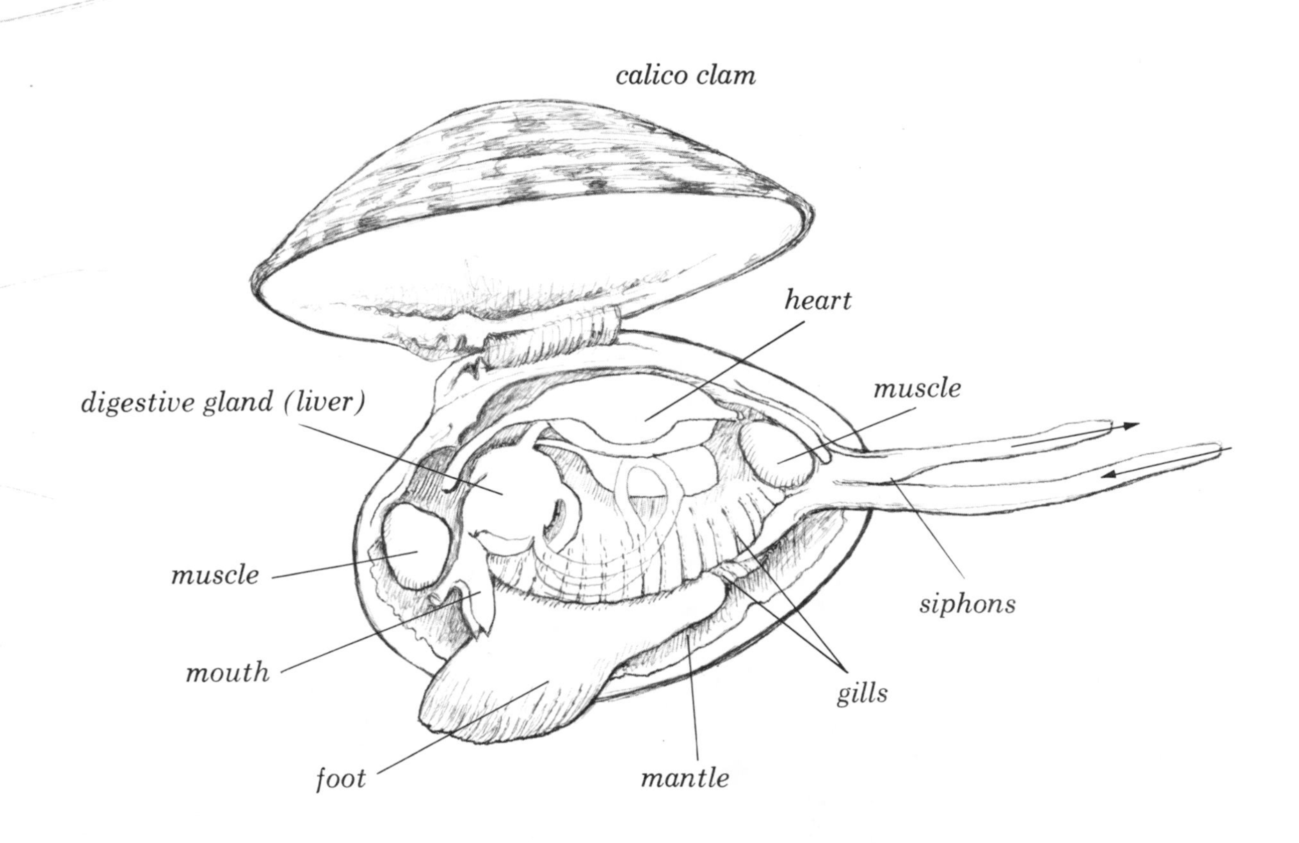

When danger comes they close up tight.

Even when their shells are closed, mollusks may not be safe. If a clam is in shallow water, a sea gull may pick it up. The gull flies high with it, and then drops it on a rock. If the shell breaks, the sea gull can get the animal that's inside. A starfish may wrap itself around a bivalve. The starfish pulls and pulls until the shells open. Then it eats the mollusk inside.

Littorina pulchra

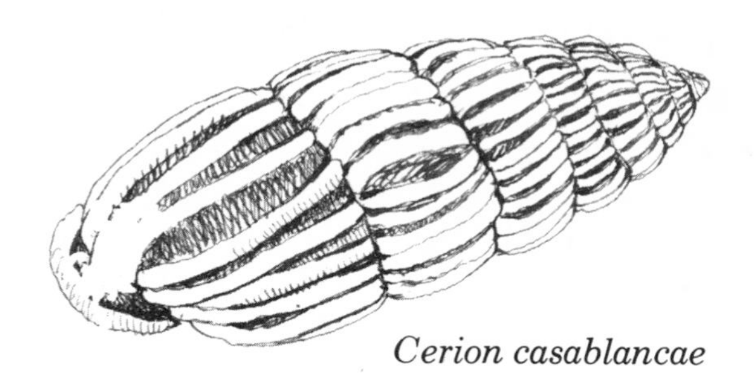

Cerion casablancae

Some kinds of mollusks are very small. Others are very big. Some live in water and some on land. There may be little snails on your garden plants. They are full-grown, but you can hardly see them.

Cepolis ovumreguli

Achatinella
Liquus
Cochlostyla
Englandina
Polymita

Clamshells may get to be as big across as a dinner plate or even bigger. In the South Seas there are clams so big you could not lift one. They could trap a careless person and hold him. Divers must be very careful to keep their hands and feet out of these giant clams. They are so strong that once the valves are closed no one can force them open.

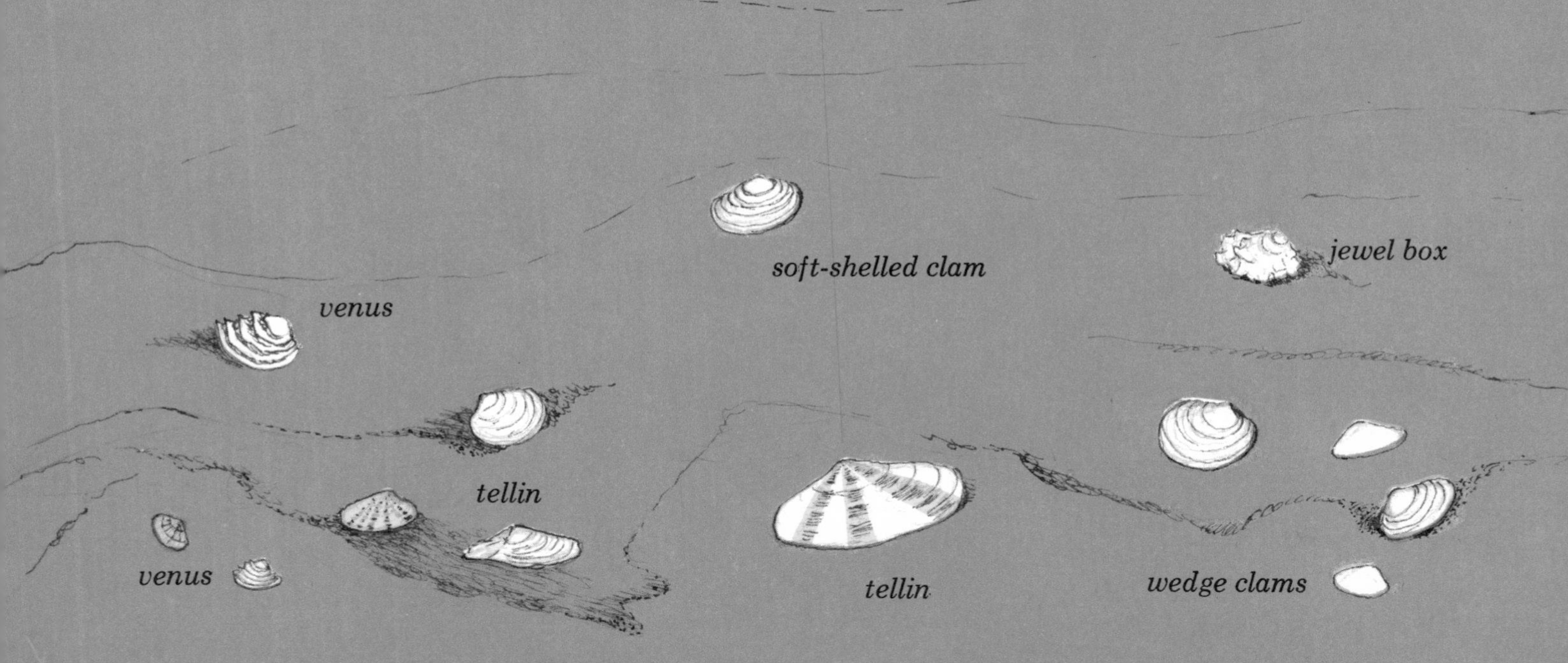

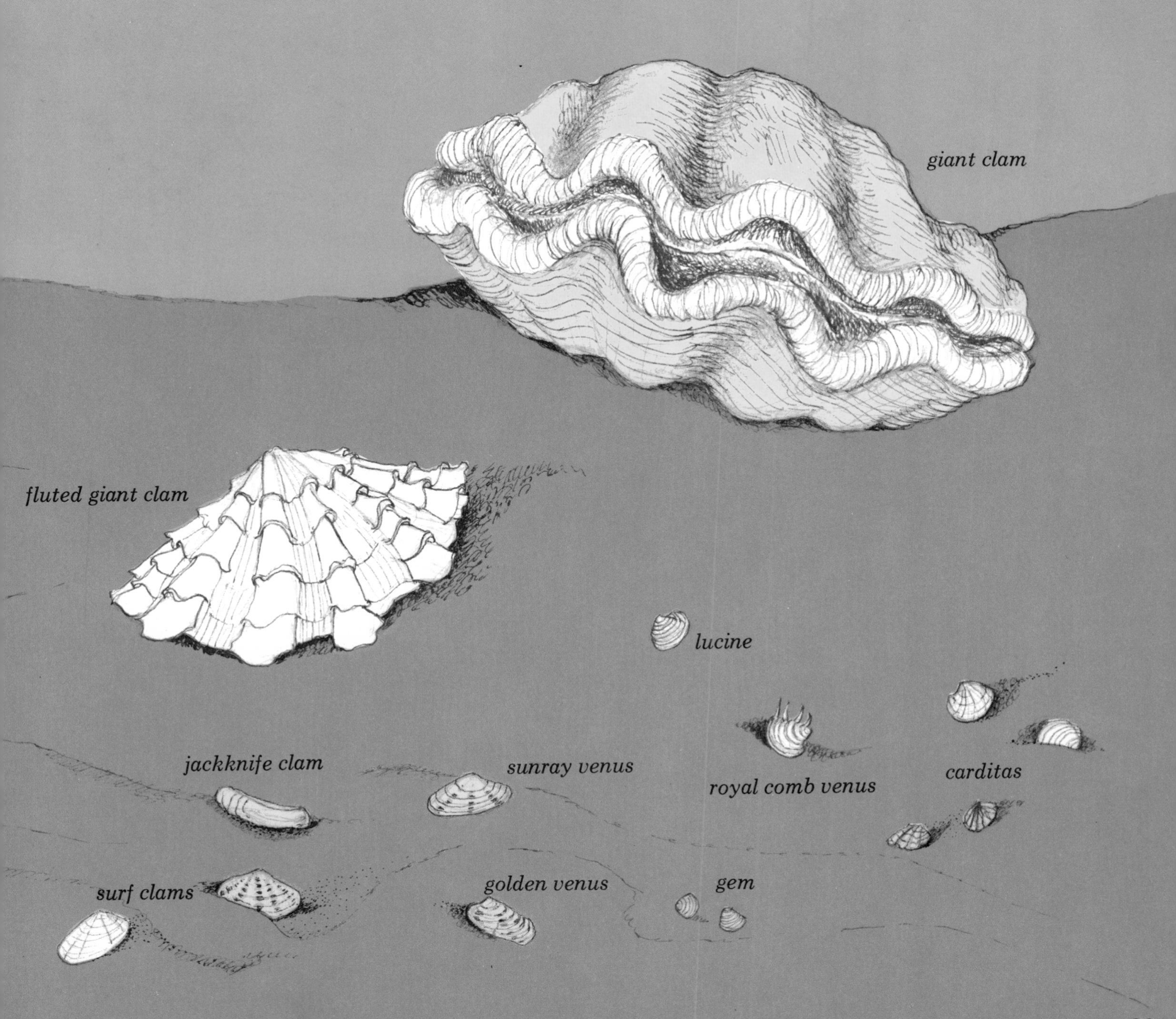
giant clam
fluted giant clam
lucine
jackknife clam
sunray venus
royal comb venus
carditas
surf clams
golden venus
gem

Most mollusks live a short time. Some scallops may live for only two years. Other mollusks live much longer. Some oysters live about ten years. And some mussels may live for thirty years.

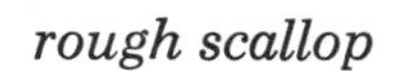

rough scallop

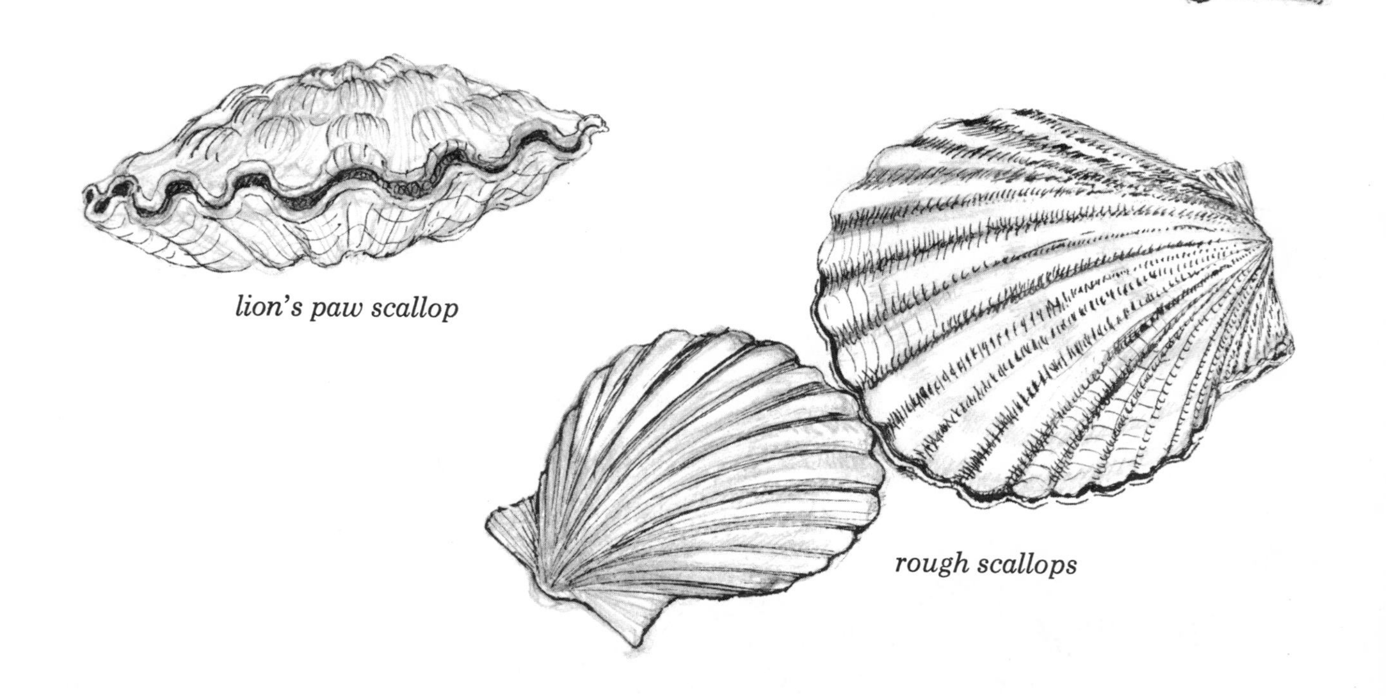

lion's paw scallop

rough scallops

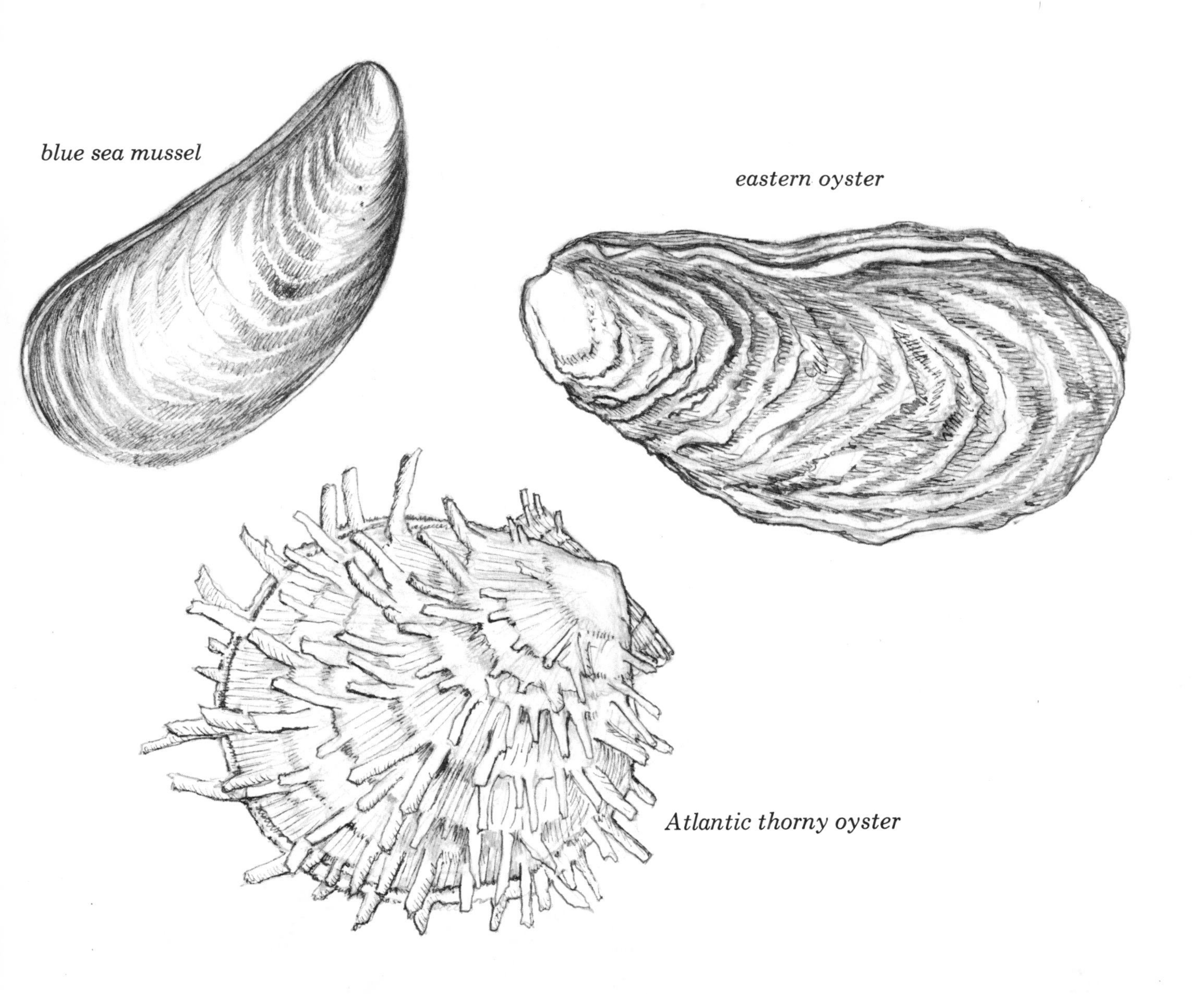
blue sea mussel
eastern oyster
Atlantic thorny oyster

If you are at the seashore, you may find many different kinds of shells. But you can find shells in other places, too. Look along the edges of ponds, lakes, and streams. You may find freshwater snails in fish ponds. In damp woods you may find land snails. You will see that shells come in many different sizes and in different shapes and colors. All of them are skeletons that were worn on the outside.

About the Author-Illustrator

As an enthusiastic snorkeler and scuba-diver, Joan Victor first became fascinated by the beauty of shells and then went on to study them scientifically.

Ms. Victor was graduated from Sophie Newcomb College of Tulane University, received her M.F.A. degree from Yale University, and has studied art therapy at Bank Street College in New York City, where she now lives. She is the illustrator of many books for young readers, a number of which she has written herself. In addition to her work as an artist, Ms. Victor enjoys swimming, biking, photography, a variety of crafts—and her favorite hobby of all, her two children, Daniel and Elizabeth.